MINING IS THE
FUTURE

Three-Step Process to Achieve
Sustainability in Mining

ALP BORA

◆ FriesenPress

One Printers Way
Altona, MB R0G 0B0
Canada

www.friesenpress.com

ISBN
978-1-03-916356-0 (Hardcover)
978-1-03-916355-3 (Paperback)
978-1-03-916357-7 (eBook)

1. BUSINESS & ECONOMICS, INDUSTRIES, NATURAL RESOURCE EXTRACTION

Distributed to the trade by The Ingram Book Company

To all sustainability professionals and advocates
within the extractive industry

CONTENTS

FOREWORD BY
PROF. VIKRAM G. YADAV

Visualize the scene that I am about to describe to you. You wake up in the morning and gaze at your window. The weather forecast and the day's agenda appear on the glass. You step into the shower and bathe in water whose temperature has been precisely calibrated to suit your body using carbon-neutral energy. As the warm water from the shower drains away, its heat is captured using a thermoelectric generator and re-used to make your morning coffee. You are eventually driven to work in an autonomous, electric vehicle. You video-call your mother and then catch up with the day's headlines. Commuting is finally enjoyable. The roads are quiet, and pedestrians are enjoying the symphony of the morning birds. There is no air pollution. People are happier and healthier.

These visions of a better future are impossible to comprehend without mining. For instance, the transportation sector is one of the largest emitters of carbon dioxide; completely electrifying the sector could reduce these emissions by over 50 percent. The transition to a completely electrified transportation sector will be built on an abundant supply of copper, nickel, cobalt, lithium, gold, silver, and rare earth metals. The most revolutionary battery technologies for energy storage are inconceivable without metals and metal oxides. Green chemical manufacturing processes—which promise to deliver the materials of everyday living more sustainably, economically, and by eliminating pollution—also employ novel catalysts that are predominantly metal based.

Yet the notion that humanity's path toward net-zero will be largely paved by the mining industry seems inherently paradoxical. The mining industry itself ranks among the largest emitters of carbon dioxide and generators of waste. An average-sized open-pit mine that annually yields 3 million tons of ore will also produce 1.5 billion tons of waste rock and 50 billion liters of wastewater called tailings. While fugitive coalbed methane and power consumption account for a large share of carbon emissions at mines, the waste rock can also lead to the formation of a highly acidic stream of liquid waste called Acid Rock Drainage (ARD). Some mines can generate as much as 1 billion liters of ARD each year. The mining industry currently accounts for roughly 7 percent of global carbon emissions. The global inventory of mine tailings and ARD totals 50 trillion liters. If the mining industry were to attempt to treat and manage this enormous volume of waste using current technologies, the total emissions by the sector would increase to 10 percent of global emissions, not to mention the steep costs. At these rates, transitioning to an entirely electric transportation sector may actually increase carbon emissions, not decrease it!

The mining industry needs logistical and technological innovation if it is to deliver the future that we desire for our future generations. To their credit, mining companies have implemented aggressive reforms to decarbonize their operations and have invested heavily in operational transformation to improve climate resiliency, enhance safety, and reduce their dependence on fossil fuels and demand for water. Nevertheless, a lot more still needs to happen. Innovations in mineral extraction, processing, and waste management could lead to even more aggressive reductions in carbon emissions and reduce the environmental footprint of the sector. However, implementing change in the mining sector is hard, largely due to the formidable scale of operations. New models for scaling innovation are needed, as well as new mechanisms for investment and increased collaboration between stakeholders such as mining companies, universities, national laboratories, startups, and regulators.

The biggest difference between a futurist and a technologist is that the former is a dreamer, and the latter understands thermodynamics. Individuals who combine the traits of a futurist and a technologist are called visionaries. Alp Bora is a visionary, and it has been a privilege to exchange ideas and strategize about the future of mining with him. He is part of a small group of thinkers whose professional journeys have included stops in process engineering, sustainability, venture capital, and policy. Not only is he deeply knowledgeable about gaps in the innovation landscape and technical requirements for inventing new technologies or improving existing ones, but he also has a finger on the pulse of investment and knows how to incubate technologies. It is a rare combination, and this book is a breath of fresh air. It is a must-read for all stakeholders of the mining sector because it breaks away from convention and introduces ideas that could transform the mining industry from being a necessary evil to one that leads by example in the fight against climate change.

Prof. Vikram G. Yadav
Entrepreneur, Innovator, Educator, Engineer
The University of British Columbia

FOREWORD BY
FRANCISCO A. CONSUEGRA

In the early 1900s, Leo Tolstoy wrote: "Everyone thinks of changing the world, but not of changing themselves." Nowadays, personal development—the act of working on ourselves to improve our life standards, performance at work, and capacity to make our dreams a reality—is widely practiced by professionals, students, and entrepreneurs alike. This opens new possibilities for societal evolution as individuals now embrace a mindset of constant change with a clear sense of purpose, which translates to increasing environmental awareness, mutual respect, and interest in contributing to society in general.

Though it's hard to trace the effects of personal development on corporate cultures and behaviors, in the past two decades, we have seen an increase in corporations investing in establishing values and—through personal development and leadership coaching—providing their employees with the tools necessary to execute their jobs under the framework of such values. This helps corporations, as their managers gain the courage to walk the talk by ensuring their actions and decisions align with corporate values.

How can we employ this lens when examining the mining industry and its interactions with our wider society? Take, for example, the concept of the IBA—the Impacts and Benefits Agreement—in Canada. Though the government holds ownership of most mineral resources, such absolute ownership is more and more contested by First Nations Communities (FNC) who have never ceded their territories and, as owners of these lands, have a right to decide both how mining is carried out in their courtyard and to claim revenues from that mining activity. This applies pressure on mining corporations, as they risk damaging their relationship with local communities if they do not maintain a healthy social contract with said communities. Consequently, mining corporations began entering into agreements with FNCs, giving FNCs access to a stable financial income as well as granting them considerable autonomy in the allocation and investment of mining-based resources, both of which help secure sustainable development for those FNCs. It has been reported that in the early 2010s, approximately two hundred IBAs were in place; ten years later, that number has increased to almost four hundred.

However, one can easily argue that these changes are the result of legal and financial pressures on mining corporations. The statistics mentioned above also imply that there is a will on behalf of mining corporations to move in such a direction. Change has not been fast, and the intentions might not have always been transparent between the two parties, but encouraging greater awareness among the leadership within mining corporations has certainly been key to enacting these changes. Such awareness can be attributed, in part, to our capacity for changing ourselves, hence the relevance of personal development.

In this quest to increase social awareness, the mining industry has, in recent years, been seeking the support of individuals who possess a deeper

knowledge of the culture of the communities surrounding mining sites/areas. This stems from government and non-government agencies alike putting
pressure on the industry and demanding corporate
citizenship to step up and do the right thing. Indeed,
nowadays it is common practice for mining corporations to have communities and social performance
teams within their senior leadership ranks that steer
their decisions and strategies in a way that enhances
relationships between the corporations and their
hosting communities. In this way, mining corporations have acknowledged and acted upon their
unspoken responsibility to leave behind sustainable
and autonomous communities.

In conclusion, the past two decades have seen the
mining companies increase their awareness regarding their role in aiding/developing host communities and the environment. This may have arisen from
a desire to mitigate the risk of losing their social
license to operate, as well as from a need to quell bad
press and alleviate pressure from host communities,
governments, and NGOs, but it also represents the
mining industry as a whole, embracing a certain level
of maturity and promoting genuine leaders who

wish to contribute to the betterment of our society and environment.

This is an ongoing process, however, and we need to understand what it will take for these efforts to succeed. In its simplest terms, we need to raise the general population's awareness of the mining industry and its contribution to our day-to-day lives.

The book that you are about to read explains in depth the reasoning behind this perspective and provides a framework for us to focus our efforts. In essence, it's an invitation for us to embrace the mining industry and give it the legitimacy it deserves as the foundation of today's society.

Francisco A. Consuegra
Mining Advocate

INTRODUCTION

Mining is a resource-hungry industry. Its consumption of diesel, natural gas, water, and other natural resources stamps a heavy footprint upon our environment. Indeed, mining as a whole is currently responsible for four to seven percent of global greenhouse gas (GHG) emissions.[1] Therefore, renewable technologies, such as solar and wind power, might seem to be incompatible with mining at first glance. And yet, renewable technology *requires* mining. Many, if not most, of the materials essential to these renewable technologies are mined. Electric cars, for example, require lithium, cobalt, and copper in levels far beyond those required in conventional cars;[2] solar panels require a range of mined minerals such as cadmium, gallium, and selenium (among others);[3] even nuclear power, often touted as a replacement for fossil fuels, needs extracted metals in the form of uranium, zirconium, hafnium, and thorium.[4]

This leads us to a paradox: if the mining industry is not clean, how can we call renewable technology clean? If large amounts of carbon emissions are created during the extraction process of materials that will go on to be used to create devices designed to reduce carbon emissions, have we truly reduced our carbon footprint?

If the mining industry is not clean, how can we call renewable technology clean?

If we wish to create a greener future, mining must change. Its carbon footprint must be reduced, its environmental impact must be diminished, and sustainability as a mindset in mining must be embraced. My goal in this book is to make the case that this is the only responsible path forward.

Mining suffers from a public opinion crisis. This is nothing new. Since the arrival of Corporate Social

Responsibility (CSR) in the late twentieth and early twenty-first centuries, mining has been one of the most heavily criticized industries for its lack of social responsibility. A criticism that was not always without reason, to be fair. For a time, however, mining was able to weather the storm. This is no longer the case. As the age of the average mining worker climbs toward retirement and demands for socially conscious investing among the labor pool grows ever more strident, mining faces a problem: how can the industry continue operations on the scale necessary to meet ever-increasing demands for minerals while fewer and fewer people choose mining as a career or, indeed, even remain aware of mining in general?

The mining industry should see this as an opportunity, not as a punishment. As Environmental, Social, and Governance (ESG) values gain importance among investors and the labor force, embracing sustainability can only aid the industry by drawing in more capital investment and a larger, more diverse workforce. And, as mentioned above, sustainability and mining are already intimately connected due to renewable technology's dependence on mining for its own production. Exploring and improving that connection will secure mining's future.

I first became truly aware of this issue after I moved from a rural mining town to a cosmopolitan urban area. As I met new people, small talk drifted toward careers, vocations, and other assorted money-making endeavors. Time and time again, people asked me what I do, after divulging, with varying degrees of humility, their own professions as lawyers, doctors, lecturers, and so on. I always gave the same answer: I work in mining.

I said this with enthusiasm because I was proud of my job. While the industry was by no means perfect, I knew its contributions to society were considerable. Foremost among them, of course, was that we provided the raw materials for fighting climate change. I knew this meant that as we moved toward a greener, more environmentally friendly future, we would also move toward a future where mining's outputs would be needed more than ever. And, as a professional in the mining industry, I was helping to build this future.

Of course, I did not say all this, but I figured people might at least be interested, as mining was not a common career choice in that city. I received a number of different responses: mostly blank stares, vague nods, and—on one memorable occasion—genuine interest, followed shortly there-after by the statement: "That's so cool! So, like, for Bitcoin, or Ethereum?"

It was an eye-opener. I'd lived and breathed mining for so long, I'd never really considered people might know so little about it. I admit, I'd expected some stronger reactions—faint distaste maybe, or even outright disdain, perhaps based on environ-mental grounds. But this apathy came as a shock. In many ways, I saw it as a worse outcome than hostil-ity. A powerfully negative opinion is still an opinion, after all.

Each time, I came away from these interactions more and more concerned. In the remote mining town where I'd lived prior to moving to the city, these reactions would have been considered ludi-crous. Everyone knew what mining was, what it was for, and why it was essential. That was the mining-town wisdom I had known, a belief system perhaps best exemplified by the saying "if it's not farmed, it's mined."

But the people I interacted with in the big city did not seem to know, care, or even think about mining. This confused me. How did they imagine all their valued electronics were made? From where did they believe the minerals and metals composing our everyday utilities originated?

More and more, I realized they didn't think of these things at all. They knew that Androids and iPhones were things assembled in factories far, far away. That was the extent of their knowledge. Where those factories sourced their materials, and how those materials were extracted from the earth, were questions not thought of, let alone asked.

This troubled me for several reasons. Foremost among them was the age of the people with whom I discussed mining. These people were, for the most part, young professionals. And considering the number of people I met, I think they represented (other than their age) a decent cross-section of society. I felt like I was single-handedly surveying a focus group. Grim findings, regardless. If young people were this disinterested in mining, it spelled trouble for the industry as a whole. Not now, perhaps. Not next year, maybe not even the next decade. But the decade after? And then the one after that? Mining, despite its popular association with brute physical labor—work often considered best suited for young people, for any number of reasons—tends to attract an older labor pool, not a younger one. The average worker in the mining industry is over forty.[5] That isn't a problem now, but as that workforce heads into retirement, the outlook becomes increasingly grim. We will need younger people as replacements, and sooner rather than later.

But what bothered me even more was their pure ignorance of the effects of mining. These were people who lived and died by their technology: they worked every day from laptops, tablets, and smartphones. They rode to work in cars, trucks, buses, and trains. All these things *need* mining because they're all composed of metals and minerals. And these metals and minerals do not simply appear from thin air. They must be mined. It's that simple; there really is no other option. Their lives—*our* lives, for I am no different in terms of how I work—depend on mining. Everyone I spoke with couldn't care less. And yet, I found I could not blame them. While the mining industry has done a lot of great work in recent decades in terms of community outreach, the public narrative is still dominated by out-of-date views that paint the industry in a negative light. What mining needed to do was tell its stories in an engaging and relevant manner; but as far as I could tell, that was not a common practice within the industry. That is what I want to change.

Mining is an "out of sight, out of mind" industry for the majority of the population, and even more so for the younger generations. Solving this issue will take time, consideration, and—above all—a plan.

Mining's Public Image Issue

In the past two decades, mining has suffered from a public image problem. Too often, mining appears in the news only in the context of one disaster or another: a cave-in, a tailing dam failure, an explosion, or yet another dispute with Indigenous peoples over mining rights. As an industry, the public pays it little attention other than these occasional stories of both rare and newsworthy events. There are reasons for this, naturally. Many of them are true: mining is a hazardous industry. However, if the public only hears negative news about mining, this can cause a vicious cycle as members of the mining industry, aware of their reputation, consider reforming it a lost cause. Propagating some of the more positive mining-related news may not only help improve mining's reputation among the general population, but might also help improve the industry's perception of itself.

Until that happens, mining will continue to suffer from its reputation as a so-called "dirty" industry. The very word connotes industrial robber-barons strip-mining the natural world, rending great holes in the earth that belch sulfurous flames. Depictions of mining in pop culture lean heavily into this imagery. Even something as fantastical as the famed *Lord of the Rings* movie trilogy from the early 2000s includes a mine as a sort of dungeon, the miners murdered by the creatures they unearthed. Such imagery is pure fantasy, constructed as part of a rousing tale for the enjoyment of filmgoers. Yet at the same time, such imagery sticks with the public. Particularly the younger generations, for whom the Mines of Moria might be the first (in many cases, only) depiction of a mine they've ever encountered.

Such poor public perception may indicate an industry on the decline in today's fast-paced, social-media-driven world. In some ways, this is far from the truth; yet in others, it strikes a bit too close to home. The supply chain for cobalt is an example of both mining's failures and its successes. Cobalt is used in lithium batteries, which power our smart-phones, laptops, and electric vehicles, among other things. As demand for these technologies exploded throughout the last decade, the demand for cobalt likewise skyrocketed.

Sixty percent of the world's cobalt supply comes from the Democratic Republic of Congo (DRC), where much of the population suffers from extreme poverty; the economic boost provided by mining could, therefore, be a boon to the country at large. Unfortunately, many workers in the DRC, including children, end up working in harsh and dangerous conditions to meet this increased demand for cobalt. That the mining industry benefitted from and encouraged such exploitation is a shameful thing, and the exact sort of practice we need to challenge and reverse whenever we can.

Thankfully, this practice was challenged in 2019 via a lawsuit filed in Washington, DC, by a human rights firm on behalf of fourteen parents and children from the DRC. The lawsuit was filed against Apple, Google, Dell, Microsoft, and Tesla. This was a landmark case, in no small part because it acknowledged that obtaining raw materials at any cost—either human or environmental—was no longer acceptable.

The situation in the DRC is far from resolved. It will take more than a lawsuit filed in the United States to fully address the issues with labor exploitation in that region. But nevertheless, the lawsuit was an important milestone, as it demonstrated that the consumer—which we all are, one way or the other—has the power to hold large corporations accountable for their global impact. Mining must acknowledge this shift in power dynamics and collaborate with its consumers in the future to reduce both the human and the environmental cost of extraction.

Mining in a Post-Pandemic World

Fortunately, there has never been a better time to pursue this goal. Mining is one of the few industries that emerged from the global COVID-19 pandemic in strong financial and operational shape.[6] At a time when so many other industries suffered setbacks, this is an unvarnished success. However, it will be remembered only as a short-term victory unless the mining industry as a whole makes the necessary macro-level changes.

This brings us back to the adoption of ESG-centered values. Mining must undergo a fundamental transformation in mindset on par with the paradigm shift that accompanied the adoption of safety standards throughout the industry in the twentieth century. With the security afforded by its currently strong financial and operational position, the mining sector has the rare opportunity to pause and reimagine its operations into a greener, more sustainable industry while the world is realigning to post-pandemic norms.

Sustainability as an industry buzzword is nothing new. Like many buzzwords, it is at its core a sound concept that has evolved over time into something that can be interpreted in countless ways. Sustainability is best understood as "meeting our own needs today without compromising the ability of future generations to meet their own needs."[7]

That's a simple definition for a complex concept. For now, it will suffice. We can explore it in greater depth later, as we build upon the specifics of sustainability in mining.

The big question behind all of this is: "Why sustainability?" And, perhaps more distinctly: "Why now?"

The answer is threefold: the future-proofing of labor in mining as an industry; ensuring continued capital investment in mining companies (especially early-stage); and maximizing long-term, socially responsible, environmentally sustainable profits for mining companies.

The AIM Framework

I call this approach taking *AIM:*

www.alpborainc.com

Awareness of the Impact of Mining

We need to spread the good word about mining. Mining needs to tell its story and let the world—especially its younger population—know its purpose, and why and how it does what it does. The mining industry need to share its stories with the population at large, not just at conventions and over podcasts aimed at an industry-specific audience. We must reach out to our children, to high school students, and to college attendees. Some of these stories may not be immediately apparent, but they still need to be told.

> **Mining needs to tell its story and let the world—especially its younger population—know its purpose, and why and how it does what it does.**

For example, I have a friend who recently announced his engagement. His fiancée is a lovely, kind-hearted woman. She worked as a therapist for traumatized children at a local hospital. It was a stressful job, but one at which she excelled. She started experiencing some abdominal pain in the months after their engagement. At first, she wrote it off as the effects of stress and worry. Not optimal, perhaps, but not exactly unexpected in her line of work. However, the pain and discomfort heightened over time. Finally, it reached a point where the "grin-and-bear-it" strategy no longer sufficed. She went to the doctor, but received scant explanation. They referred her to a specialist. There, she underwent some procedures, including a CT scan. They found an abnormality on one of her internal organs. It was nothing life-threatening for now, but it had a strong chance of causing cancerous growth later in life. Within weeks, she underwent laparoscopic surgery. It was a success. The abnormality was removed, she recovered over the following months, and she and my friend continue their lives together to this day, happy and cancer free.

At first, the connection to mining in this story might seem tenuous. But consider the following: at all points within her medical journey, this woman and her doctors required the aid of several specific medical technologies. The CT scanner to identify her illness, the laparoscopy machine to fix it, and various monitoring technologies during and after the procedure to ensure she recovered. Without these technologies, her ailment could not have even been identified, let alone cured. Each and every one of these technologies depends on metals and minerals that were extracted by mining: lithium, copper, gold, and so on. All are essential parts of their batteries and circuitry. Lithium is found in pacemakers and defibrillators, for example, while titanium is essential for surgical pins and bone plates.[8] Without mining, such technologies would not exist. Therefore, without mining, modern medicine (and all the people it helps treat) would not exist.

These are the stories mining advocates need to tell: ones emphasizing the interconnectedness of mining to our lives and those industries that maintain our lives.

Agriculture has accomplished this goal to some extent. Consider the growth of farm-to-table restaurants, farmers' markets and shop/eat local movements. Everyone eats, and the agricultural industry has made the most of this fact by embracing local, sustainable outreach, while earning a new image in the public eye in the process.

If you eat, you are involved in agriculture.

Mining needs this same outreach. While people can't necessarily pair their backyard chickens and herb gardens with a backyard copper mine, the awareness outreach still matters. People place considerable weight on the provenance of their meat and produce;

imagine if they did the same for their metals and minerals. It would be a revolutionary development.

Investment in Innovation, Technology, and the New Generation

If mining is to continue to thrive in the coming decades, it needs to invest in attracting a new generation of workers. This can mean literally a new generation—millennials and Generation Z, for example—or it can mean new demographics and a more diverse workforce.

Currently, mining is still dominated by men, which is not inherently an issue (there are female-dominated industries as well, such as education and nursing), but it does limit the industry when it comes to attracting new workers. Millennials and Generation Z will constitute the vast majority of the available labor pool within the decade, while women continue to be both a slim majority of the general population as well as the majority of the available labor pool. If mining continues to grow, it will need new workers, and these new workers must come from one of the above demographics. Therefore, mining must brand itself in an appealing manner to attract a larger and more diverse segment of the population.

In addition, mining (like all industries) requires capital investment. For mining in the United States in particular, money must flow toward the industry, as the US is entirely dependent on imports for fourteen out of the thirty minerals critical for maintaining and advancing our modern infrastructure. Moreover, the US is importing about 50 percent of its demand for ten critical minerals. If we wish to maintain our infrastructure while also reducing its impact on the environment, we must invest in domestic solutions for the extraction of these minerals. Mining—especially early-stage mining companies—must attract large, institutional investors if we wish to change these supply-chain dynamics.

Mindset Shift Toward Sustainability

This brings us to our final point. Our ultimate goal is the transformation of mining from a perceived "dirty" industry into a new, green industry. If we wish to accomplish this while also attracting a vibrant, motivated, and younger workforce, then mining must shift its mindset toward sustainability. We need to position ourselves as a renewed, reenergized industry whose aim is the production of metals and minerals for the benefit of all humankind. This needs to be an organic repositioning of our mindset, focused on every level of worker from the ground up. In other words, it should not just be a problem for the environmental team or the PR department. We must balance purpose and profit, for one without the other is simply not viable in the long term. Social responsibility must hold the same weight as our profit margins. In today's atmosphere, such an outlook will benefit us all. A high tide lifts all ships, after all.

> **Social responsibility must hold the same weight as our profit margins.**

Indeed, a focus on sustainability aids both investment and awareness among the new generation. Younger and more diverse workers increasingly place sustainability among their core values. In addition to impacting political and social matters, it affects which careers they pursue, even which companies they choose. An industry-wide paradigm shift toward sustainability kills two birds with one stone: it raises awareness while also investing in the future generation of mining workers.

A shift toward sustainability involves undergoing a learning process. Like so many processes, it can be broken down into steps. Each step builds upon the next, and thereby enhances the end product. The process as a whole is more than the sum of its parts. In the following chapters, I'll explore in detail how each step in this process might look, and why this transformation is essential for the future of mining as an industry.

CHAPTER 1

AWARENESS

The visibility of mining as an industry is increasingly diluted by misinformation, despite its increasing relevance to society at large. While advances in technology have improved our ability to communicate, recreate, and innovate, many of the high-tech devices used for these purposes rely on materials produced by mining. From the rare earth minerals which are essential to an iPhone to the silver in pollution-reducing catalytic converters, mining plays a key role in enabling advanced technologies—and this is especially true for environmentally friendly applications.

Yet almost all mining takes place in physically remote, economically depressed regions far from the public eye. This distance provides fertile ground for misinformation. Mining workers are often characterized in popular culture as grubby, poor men pickaxing metals from the earth while simultaneously suffering at the hands of monocled robber barons. This image owes more to inherited nineteenth-century stereotypes than to current realities. Consider this example from popular culture: the science-fictional movie *Avatar* portrayed its mining company as despoilers of the environment and slayers of an Indigenous culture, propagating the stereotype of mining as an exploitative and immoral industry. Even something as innocuous as *Snow White* shows the dwarfs as stereotypical miners with pickaxes, rooting the industry in a primitive imagery that ignores its modern advancements. Dispelling these images and increasing mining's visibility as a vibrant, modern industry—and thereby attracting a vibrant, modern labor force—means increasing the general public's awareness of mining. However, accomplishing that means changing the industry's current methods of communication. Right now, success stories in mining are published as LinkedIn posts, whereas its failures earn screen time as national news. Altering that dynamic is the essence of building greater awareness.

> **Success stories in mining are published as LinkedIn posts, whereas its failures earn screen time as national news.**

So, what *is* awareness, then?

Consider this: I recently bought a car. While doing my typical research rounds before purchase, I noticed a strange phenomenon. Once a car (or rather, a type of car) catches my fancy, I find myself noticing it out in the world more often. If I'm looking at a Tesla, or a Volkswagen Beetle, or even something as titanic as an F150, it's the same: once my mind has been opened to the idea of desiring that car or truck, I see it everywhere—passing me on the highway, parked at a restaurant, or sitting in a driveway. I become intimately, inescapably aware of this vehicle model. From a rational standpoint, I know there's no way my mental state changes the actual prevalence of these vehicles in the real world. There were, after all, the same number of Teslas and F150s on the road before I did my research as there were after I completed my research. And yet, my perception remains the same.

So, what changes? Why do I become more aware of something that has always existed once I've researched it? Why does my state of mind change my experience of the outside world?

There is a lot of fascinating psychology and philosophy one might unearth from those questions. But for our purposes, only one answer need suffice: the Baader-Meinhof phenomenon. A strange name for a familiar concept, the phrase was coined after an incident where a certain man was discussing the Baader-Meinhof gang (a group of communist terrorists in Cold War Germany) decades after their relevance to popular culture had passed. After learning about this somewhat obscure historical tidbit, he found himself suddenly noticing references to this group everywhere: a news article on past members, a reference to their acts in a favorite movie, mention of them in song lyrics, and so on. Baffled by this, the man shared his new awareness on an Internet discussion board. Others reported similar experiences, and thus a new term was born.

This phenomenon is an example of what Stanford linguist Andrew Zwicky called "frequency illusion."[9] In other words, once we become aware of something, we are far more likely to notice it and comment on it and, indeed, believe it more prevalent than before, despite its actual status neither increasing nor diminishing.

> **Once we become aware of something, we believe it more prevalent than before.**

This is a type of awareness—the sort of awareness that mining as an industry requires if it wants to update its image among women and younger workers. Mining needs its story to be out in the world, floating around the general community, demanding to be noticed. Indeed, that is the first step: the mining industry must tell its story. Narrative provides the impetus that transforms knowledge into action.

Narrative-Building as Awareness

Revamping mining's public narrative is key to increasing the general population's awareness of mining as an industry. Accomplishing this first requires us to understand the power narratives hold within our modern society.

I have a thought experiment I like to run through along these lines. I've mentioned above how the agricultural industry revolutionized its public image through a collective rebranding. I hope that mining will go through a similar transformation. The narrative trajectories for both industries mirror each other to a great extent, after all.

Prior to the nineteenth century, the majority of the developed world lived in close contact with agriculture. The public image of farming once focused on what we might call the "yeoman farmer": the self-employed, self-starting farmer of the rural area. This public image was not always correct, but it was influential, as so often occurs.

As the industrial revolution took hold throughout the nineteenth and early twentieth centuries, this image of the yeoman farmer lost its cultural relevance. Agriculture throughout the twentieth century became the pop-cultural home of rural farmers and corporate farms. Controversies around unethical practices, genetically modified organisms, and exploitative labor conditions seemed to have irreparably damaged the industry's image. Despite being an essential industry, agriculture's public image had never been worse.

Then, in the early twenty-first century, this trend reversed. Negative stereotypes surrounding agriculture gave way as part of a fresh social movement. A renewed culinary focus on locally sourced, locally farmed ingredients became popular in high-status cultural centers such as New York City and Los Angeles. Farmers' markets, local foods, and small-scale farming became boutique culture. The public image of agriculture was no longer tied solely to the excesses of factory farms or the classist stereotypes of the rural working class.

Now when I purchase beef, chicken, apples, or grapes at my local supermarket, each product often comes stamped with a local farm's name and location. Indeed, I've eaten produce that arrived with the names of its pickers labeled proudly on its packaging. When I eat at local restaurants, the menus display each item's farm of origin in bold lettering. All these facets together add up to an awakening that transcends the individual and attains cultural status. Though I was never ignorant of agriculture, I did not think very much about it. Food was something purchased from a grocery store, or sometimes obtained from back-country relatives with a fondness for bowhunting. Now I know the names of local farms and farmers. Whereas I'd have struggled to name one farm in the past, I now prefer to purchase from specific farms. Nor am I alone in this. It's a nationwide—worldwide, at this point—phenomenon.

Such is the power of awareness and narrative. My thought experiment runs thus: imagine that level of awareness turned toward mining. While it might be easier for the average person to visit a local dairy farm than it is for them to visit a local mine, the general principles still apply. Knowing the provenance of one's food lends a sense of identity to the meat, grain, and produce consumed. That sense of identity is powerful, particularly for members of our younger generations. It gives a sense of connection to something they're already consuming—and indeed will always need to consume. A personal sense of identity and connection provides a strong, motivating force.

Mining is no less essential to our modern lives than agriculture. The ubiquity of technology in modern life means all of us are intimately connected to the fruits of mining, whether we realize it or not. And just as agriculture has managed to do, we need to connect people with their mining materials in this intimate, personal way.

Stories can provide this connection—or rather, *positive* stories. We need stories about the ways in which mining changes the lives of its workers for the better. What's more, we need stories demonstrating the myriad ways that mining brings about positive change, both directly and indirectly.

> **We need stories demonstrating the myriad ways that mining brings about positive change, both directly and indirectly.**

Currently, certain narratives dominate mining. For example, I recently searched for books about mining on Amazon Books, which turned up seventy-five pages of results. Each page listed about twenty books. At first, I was impressed. Daunted, even. How to separate the wheat from the chaff (or the ore from the waste, rather)? Sifting through seventy-five pages of material for the right book seemed something of a tall order.

I needn't have worried. On that first page alone, there were precisely two books on actual, minerals-extracted-from-the-ground mining. One was an introductory college textbook with the rather literal title, *What Is Mining?* The second was a nonfiction account of an infamous mining disaster from the nineteenth century. The other eighteen results? Works on data mining, crypto mining, and other such applications. When I went to Goodreads, the largest book review site on the web, I found similar results.

The results were clear. The narrative surrounding mining was either nonexistent or negative. That needs to change. Easier said than done, of course. For agriculture, the process took decades. Indeed, the process continues to this day, albeit with much of the work accomplished already. Unfortunately, mining does not have the leisure of such a lengthy schedule. Today's world moves too fast, and mining's demographic issues are too immediate. Fortunately, mining possesses an advantage over the agricultural industry.

Sustainability as Mining's Narrative

That main advantage stems from an unlikely corner: sustainability. Sustainability forms one corner of that familiar triad, ESG. In this context, ESG refers to a method of more rigidly defining socially responsible investing, where corporations rebrand themselves as investors in the welfare of society at large rather than heartless entities focusing solely on profit at all costs. Currently, there is a consumer-driven boom in sustainability demand. Businesses may not drive this demand, but they must respond to it.

There is a consumer-driven boom in sustainability demand.

Mining is no stranger to the concept of ESG. In the early twenty-first century, mining was a pioneer in producing environmental impact studies, which are an important tool in analyzing and assessing a corporation's ESG score. Of course, these studies were generally government mandated, as extraction industries such as mining and petroleum were (and are) considered to have a high impact on the environment. Regardless of its origins, ESG has been a part of mining's makeup for decades. We already have the tools and the infrastructure to take advantage of this new trend.

ESG has been a part of mining's makeup for decades.

Unfortunately, within the industry itself, ESG may carry negative connotations. Too often, meeting sustainability criteria is considered burdensome, especially withing the frontline workers, and just a way of obtaining the "license to operate," or yet another concern the public relations (PR) department must address before the local community accepts a new mine in their area.

There is some truth behind these concerns. Often, both now and in the past, ESG concerns have been at odds with mining. But that should no longer be the case. Earlier in this book, we discussed mining's importance vis-à-vis technology. More directly, we discussed its relationship to medical technology, as well as current computing and smartphone technology. The same holds true for sustainable, renewable, and other "green" energy technologies. These technologies—whether they be wind powered, solar powered, or otherwise—rely on the minerals extracted via mining.

For example, constructing a solar panel requires nineteen minerals extracted via mining. Eight of these minerals are listed among those internationally deemed "critical minerals," meaning they're particularly important to clean energy technologies production and also face supply challenges, such as a small global market, a lack of supply diversity, market complexities caused by co-production, and geopolitical risks. Some of these critical minerals include aluminum, lithium, titanium, and uranium. Without them and other minerals like them, there would be no green technology.

Years ago, I worked at an aluminum smelter. I learned that aluminum can essentially be recycled *forever* without losing its essential properties, so long as it is recycled properly. However, if it is not recycled, then it simply becomes yet another waste product. This is heartbreaking when one considers all the work that goes into creating consumable aluminum: first it is mined as bauxite, then refined into alumina, smelted into aluminum, and finally, manufactured. All that work requires extensive resources, and much of it could be avoided if people knew that the aluminum foil they use on their cookie sheets and pizza stones is a renewable resource and not something suitable for the garbage. Aluminum cans, for example, while only representing 1.4 percent of any given ton of waste, constitute 14.1 percent of greenhouse gas emissions when one accounts for the time and energy spent replacing them via freshly mined bauxite.[10] That is why recycling and reclaiming metals like aluminum—even those used in household tasks—is so important. Personal efforts, such as utilizing recycling bins and bottle returns, can help, but only if aided by broader legislation aimed at incentivizing recycling; so-called "bottle bills," for example, can lead to the recycling of upwards of 80 percent of aluminum containers.[11] Every ounce of recycled metal is metal that we don't have to replace through extraction processes.

These narratives of sustainability must be disseminated far and wide. Emphasizing mining's connection to these minerals and its connections to clean energy will pay dividends by explicitly drawing the connection between sustainability and mining. One does not exist without the other. This positive link between ESG and mining promotes the sort of awareness we need; by doing so, we shift mining's narrative from one rooted in flawed pictures of an imperfect past toward a future-focused, green-tech-centered image of social responsibility.

Accomplishing this narrative shift will involve embracing sustainability as a core value of the mining industry. This is easier said than done, but with the proper planning and investment, the shift can occur quickly and easily. It is that second point—investment—that I will cover in the next chapter, for it is through investment in future generations of mining workers that we will accomplish the powerful, epoch-defining narrative shift I outline here.

CHAPTER 2

INVESTMENT

Awareness is not enough. Just because I notice more F150s doesn't mean I will necessarily purchase one from a dealership. I am more *aware*, yes, but awareness alone does not persuade. Persuasion requires something more concrete. It requires *investment*.

Regarding the mining industry, our investment must focus on current and future generations. These generations—millennials and Generation Z ("Zoomers" in the colloquial sense)—place great importance on ESG values in terms of where they invest their money as well as which career paths they choose. In that sense, investing in sustainability will both attract more money to mining and more workers, thereby alleviating concerns over mining's aging labor pool and the changing investment landscape.

In 2021, the International Council on Mining and Metals (ICMM)—a multinational body composed of the twenty-eight largest mining companies worldwide—promised mining would produce net-zero Scope 1 and Scope 2 greenhouse gas (GHG) emissions by 2050.[12] Scope 1 GHGs occur when organizations (i.e., corporations) directly emit GHG via boilers, furnaces, engines, and other such sources. Scope 2 GHGs are more indirect: they are associated with the on-site purchase of electricity and steam power. In other words, a Scope 1 GHG would be the burning of fossil fuels on-site, and a Scope 2 GHG would be all the electricity used in the mine itself, if that electricity is sourced from GHG-producing plants.[13] Though not prioritized by the ICMM at the time this book was written, there also exists Scope 3 GHGs, which are those GHGs produced by all sources indirectly impacted by a company's product. This includes, among other sources, a company's own customers: for example, if a mining company sells coal to an electric company and that coal ends up being consumed to produce electrical power. It is worth noting that Scope 3 GHGs often comprise the majority of GHG emissions.[14] Committing to the elimination of both Scope 1 and Scope 2 appears, on the surface, like the sort of investment the industry requires: a strong commitment to a widespread and actionable goal within an achievable timeframe.

However, if one peels back the layers, the facade crumbles. The ICMM has offered very little in the way of a concrete plan to meet this goal by 2050. Indeed, at times it has seemed like it resists offering further plans, opting instead to keep everything vague. This sort of strategy smacks of the old way of doing things for mining: viewing environmental and social concerns as a sort of extra "tax" on the mining industry, a box to be checked rather than a duty to be embraced.

In a sense, one could see this as a repetition of ingrained, cynical patterns. On the other hand, it's also an opportunity. The ICMM, despite its lack of overall planning, has at least set an end date for a goal, which is certainly an improvement over the past. How mining reaches that end goal is up for grabs, however.

Sustainability as a Necessary Investment

An honest pivot toward sustainability is the approach I champion, as opposed to a compliance-driven effort to meet the arbitrary standards set by industry collectives. Investment in sustainability can fix some of the very real issues I alluded to previously: mining's narrative failings and its demographic challenges. I've discussed the narrative component enough for now; we will return to it later. The demographic issues, however, may be more immediate and, perhaps, even more immediately solved via the embrace of sustainability.

I first noticed some of these challenges on a plane ride into a mining site in northern Canada. It was a remote location, accessible only by plane or by a thirty-hour train ride. I didn't much care for a thirty-hour-long commute, so I chose the plane. As I was waiting in line to board, I looked around, doing the sort of idle people watching we all indulge in at times. I was still new to the mining industry at this point, so perhaps that's why I didn't immediately notice that nineteen out of twenty of us were men.

And not only men—older men. I was the youngest fellow there by at least ten years, if not more.

Intrigued by this observation, I researched the issue. What I found out only confirmed my thoughts: mining has one of the oldest and most male-dominated workforces in the world. More than 50 percent of mining workers are over forty, and only about 9 percent of mining workers are women.[15] Relatively few industries match mining workers in age, and almost none match its male-dominated aspect: out of common labor industries, the only one exceeding mining in terms of male-to-female ratio is construction.

> **Mining has one of the oldest and most male-dominated workforces in the world.**

This is not inherently an issue. A male-dominated industry is not intrinsically worse than a female-dominated one. The same logic applies to an industry with an older workforce. However, when it comes to rectifying mining's public narrative, these demographics—a younger and more diversified workforce—are the ones the mining industry needs.

Sustainability exploits this window of opportunity through its appeal to the demographics the mining industry currently lacks. In general terms, sustainability-focused business philosophies such as ESG conceptualize each company as responsible for five broad stakeholders: its workers, its surrounding communities, its clients, its own shareholders, and the environment at large. Proponents for this type of investing reason that any company serving those five stakeholders is a company focused on sustainable, long-term growth, and such companies provide higher return on investment.[16]

Furthermore, ESG specifically targets investing in companies with a strong record in managing carbon emissions, water/land pollution, and green energy initiatives, among other things.[17] ESG investing has exploded in the last decade, increasing by over 42 percent since 2011, which translates to tens of trillions in USD invested.

Therefore, a market for ESG exists, and it has attracted considerable cash influx in recent years. That alone might justify mining's interest in the concept, but there are other factors at work beyond

ESG itself. Indeed, ESG is only a symptom of wider cultural shifts in the labor pool and business world.

The aging population of mining workers, as mentioned above, is a threat to the long-term viability of mining as an industry. Increasing awareness of mining's inextricable link to ESG will partially rectify this issue by increasing the appeal of mining for the two major demographics mining lacks: younger workers—both Generation Z and millennials—and more diverse workers. Both of these demographics played a massive role in the popularization of ESG as an overall investment strategy as well as a more business-neutral value system. Without these demographics and their awareness of mining, the industry cannot thrive.

When it comes to attracting new, younger workers, a value-centered approach can be invaluable. According to recent studies conducted, about nine out of ten millennials (persons born between 1981 and 1996) would consider taking a pay cut to work at a company whose values align with their own.[18] This is a dramatic cultural shift: less than one in ten baby boomers would consider taking a pay cut for the same reason.

Millennials have been at the forefront of popularizing ESG, both as an investment strategy and as an overall business philosophy. In part, this is a numbers game: millennials currently constitute the largest portion of the available labor pool, as mentioned above. And yet, the average mining worker is not a millennial. They're older: Generation X or a baby boomer. Unfortunately, the median retirement age for such workers is not terribly far above forty, especially in careers such as mining that may involve higher levels of stress, both physical and mental. As these workers retire or otherwise leave the mining industry, it will be increasingly difficult to replace them with younger workers if mining does not reflect the values of younger generations. These younger workers place a premium on values-centered businesses and, in particular, on businesses that promote sustainability.

Mining needs new, younger workers.

In a way, there are few industries better prepared to take advantage of this cultural shift than mining. Due to our aforementioned connection to green technology, we can exploit that narrative congruency and rebrand ourselves as an industry inextricable from the future of green, sustainable technology. In doing so, we can embrace the cultural shift and attract new, essential demographics to our industry.

The Necessary Investments

Sustainable mining sounds like an oxymoron. Extraction is, after all, a one-way process. Ore, once extracted from the earth, is never replenished. At least in that respect, agriculture holds an advantage over the mining industry.

There are viable ways for mining companies to invest in sustainable practices. For example, I am acquainted with an American copper producer. They use low-carbon, renewable energy in their copper extraction, which they carry out on American soil to almost exclusively supply to another American company, which uses the copper in the production of wind turbines for renewable energy. This is the sort of narrative success mining needs: a local mineral extractor using renewable technology in their extraction process, who then sells that mineral to another company so that it might find direct use in renewable technology. It perfectly illustrates the intersection of mining and renewable technology and demonstrates both how mining can do better as an industry as well as the relevance mining holds for green technology as a whole.

Unfortunately, stories like this do not get told often enough. The end users—the everyday consumers of minerals and metals—remain in the dark. This costs us some tremendous opportunities. Now, not every aspect of this story requires replication: supplying minerals only to domestic, renewable energy producers is not viable for an international industry upon which all of modern society depends. However, their example provides a model for other mining companies. This model need not be followed to the letter, but parts of it can be used to great effect. For example, using renewable technology in the extraction process which they then supply to a company that uses the minerals for producing yet more renewable technology helps reduce Scope 1, Scope 2, and Scope 3 GHG emissions.

Another, no less significant, takeaway is the link between mining and green technology as a marketing strategy. This copper producer is redefining their narrative and performing ESG as a core company value as well as a business practice. That alone is noteworthy.

Their business model provokes another discussion. Their use of green tech to produce copper that goes on to create yet more green tech is an example of a circular economy: one based on reuse and remanufacture, where material created then serves another, renewable purpose, even if the resource employed in its own creation is not itself renewable. We might compare this circular economy to the more common linear economy, where natural resources are consumed via production and manufacturing, then disposed of as waste.

A circular economy reduces waste through its self-sustaining process of reuse and remanufacture—recycling is a prominent example of this process. A circular economy encourages alternatives to chemical sources—which are often non-renewable—and thereby helps preserve our natural systems. But even in such circular economies, there exists a place for non-renewable resources. In our example, by using copper (a non-renewable mineral) in the production of wind energy (a renewable energy source), the ultimate amount of waste generated is diminished.[19]

A circular economy reduces waste through its self-sustaining process of reuse and remanufacture.

Encouraging the mining industry's participation in circular economies is an immediate and actionable step through which the industry could invest in sustainability without decreasing overall profits. Indeed, as seen above, this would in all likelihood increase profits, both due to the increased awareness among potential workers and investors, as well as the ability to reuse and remanufacture mineral resources that would otherwise become waste product.

That last point—the reuse and remanufacture of potential waste—is the final piece of this investment puzzle. For example, the mining industry could make use of already available resources and reintroduce them into production. This would be a sustainable, green tech solution as well as a profitable one. It involves a process known as urban mining.

Urban mining involves reclaiming raw materials present in waste normally sent to landfills. Conceptually, the process utilizes waste as a vital resource, making use of human-created—as opposed to earth-extracted—stocks to cater to the manufacturing demands. Thus, it fits neatly into the circular economy we discussed above: reusing and remanufacturing already mined materials present in electronics, discarded automobile parts, and so on.

Though urban mining typically deals with recovering e-waste metals—which is to say, the metals found in discarded phones, computers, televisions, and other common electronics—the term can be used for the monetization and recovery of any materials present in the waste streams, including:

- Demolition and construction waste: recovering materials such as metals, rubber, cardboard, paper, and wood
- Solid municipal waste: general recovery consisting of anything from recycling plastics and metals to composting for commercial or resale purposes
- Tires: particularly the recovery of metal and rubber from tires or other products made using rubber.

That being said, e-waste constitutes the majority of urban mining efforts, as electronic items make use of a wide variety of recoverable and recyclable metals, including a number of the critical minerals mentioned previously such as palladium, silver, and gold. In 2019, the United States produced nearly 7 million tons of e-waste, 15 percent of which was recycled. Currently, twenty-five states in the United States have laws pertaining to e-waste recycling.[20]

Urban mining, then, is a method of sustainable mineral reuse and remanufacture, a process the mining industry already has the infrastructure to handle. Its importance in the coming decades will only increase. As a method of sustainable investment, it is appealing in several significant ways.

Firstly, the era of cheap and abundant raw materials is over. Perhaps the simplest explanation for increasing support for urban mining is that our resource reserves are becoming more difficult to extract from the earth as ore. Ugo Bardi, an earth scientist at the University of Florence, argues in his 2014 publication *Extracted* that it is becoming increasingly difficult to extract whatever raw material reserves we still have. To greatly simplify his model, fossil fuels and minerals are exhaustible resources, and their extraction can be modeled as a sort of bell curve.[21] The top of that curve—its peak—represents both the point at which extraction reaches its highest level as well as the point at which extraction rates begin declining as easier-to-reach deposits are exhausted, forcing mining operations to either mine more remote locations or employ less efficient methods of extraction. These increased difficulties in extraction drive up the cost of the extraction process itself, which of course increases the price of the mineral end product. Many of these methods are also *more* detrimental to the environment—consider the controversial process of fracking for natural gas extraction—either as a direct result of the methods used, or due to the increased energy investment required to extract these mineral deposits.

Bardi's models predict there will come a time when the extraction process becomes so monetarily inefficient that it renders mining nonviable. Bardi reckons that we may be fast approaching our peak with certain valuable minerals, such as phosphate. Reusing metals from already manufactured goods helps sidestep these increasing costs and is far more environmentally friendly than extracting ores directly from the ground.

Second, recovering raw materials through e-waste could become a cost-effective process. For many years, e-waste was treated as exactly that—waste. It was dumped and forgotten. However, we now know e-waste constitutes a valuable source of reclaimed precious metals. For example, the Environmental Protection Agency (EPA) estimates that for every 1 million cellphones, we could recycle 35 thousand pounds of copper, 772 pounds of silver, 75 pounds of gold, and 33 pounds of palladium.[22] This sounds like an unreasonable number of cellphones for comparatively little value until one considers that more than 416 thousand cellphones are thrown out in the USA alone every day.[23]

Similar to how solar energy is cheaper than using fossil fuels to generate electricity, extracting metals through urban mining could become considerably cheaper than relying on classic mining if we prioritize innovation and investment in the process. For example, the EPA estimates that one ton of circuit boards, if recycled properly, might contain forty times more copper than an equivalent weight of ore.[24]

In addition, there are still numerous appliances that are not even partially recycled. Scientists collaborating via the United Nations University and International Solid Waste Association found that in 2017, the annual generation of e-waste topped out at 44.7 million metric tons annually.[25] They estimated that by 2021, this number would exceed 52 million metric tons per year, and furthermore, only 20 percent of this e-waste was destined for recycling and reclamation. That remaining 80 percent of e-waste comprises numerous untapped mines of their own, if viewed in terms of urban mining.

Why Investment Matters

Investing in sustainable mining practices, such as employing renewable technology in our extraction processes and exploring the urban mining phenomenon, will help shift the public narrative of mining away from its negative past and toward a positive future.

As mentioned earlier, younger generations of workers place a premium on the values represented by their workplace and will often refuse a career in fields that do not match their own moral commitments; the Deloitte Millennial Survey found that around 37 percent of millennials and Zoomers would refuse a job based on ethical/moral concerns, and this number skyrockets to almost half (46 percent) when the young people in question possess senior-level positions.[26] Investing in sustainable technology and mining practices speaks to the values of younger workers and ensures mining's continued viability as the current generation of mining workers age into retirement.

Finally, exploring alternative, green methods of mining—such as urban mining—helps sidestep the falling availability of extractable minerals in traditional mining sites while simultaneously aiding in the above awareness-focused endeavors. Furthermore, these mineral waste deposits in landfills and other waste sites represent an untapped source of profit for mining companies.

In the next chapter, I will explore which individual efforts each mining company may need to take to ensure this policy of investment in sustainability sticks, so to speak, and how a shift toward sustainability involves not only a shift in business practice and marketing language, but a shift in company culture and individual mindset as well.

Investing in sustainability is *the* way by which the mining industry will increase awareness among the general public.

CHAPTER 3

MINDSET SHIFT

C reating an industry-wide paradigm shift toward sustainability as a core value will require more than just a commitment to raising awareness and investing in our future: it will require a concurrent shift in the mining industry's mindset. This shift in mindset toward a sustainability-focused industry will be dramatic, but it is not unprecedented. Mining has undergone similar paradigm shifts previously, and it can do so again.

My first job in the industry took me to Sudbury, the nickel mining capital of Canada. It was a great experience. But there was a bit of culture shock when I first began my career. I came from aerospace engineering, so I was unprepared for a part of my first day at the nickel mine: the safety share. In fact, when my supervisor first summoned me for the meeting, grumbling something vague about "safety," I felt a mild tinge of fear. I figured I'd done something wrong, somehow, and they were going to drag me in front of the boss for the customary reprimand.

Turns out it was a ritual, just not the one I thought. Classic new guy jitters.

The safety share is a brief meeting performed almost every morning at almost every mining company. They can last anywhere from thirty seconds to ten minutes, during which we share a behavior or condition related to safety (at home or at work) in order to ignite awareness about safety and keep it always top of mind.

At first, this felt like a strange practice. But in time, the safety share became second nature, both for me and all the other employees.

Indeed, it was at that first safety meeting that I first heard what became a familiar refrain: "the most valuable resource coming out of any mine is the worker." For me, this was a revolutionary idea—a real, employee-centered core value that was ingrained at every level of the company. Keep the workers safe, and all of us will prosper. It made sense then, and it makes sense now.

> **The most valuable resource coming out of any mine is the worker.**

And yet, what I was experiencing did not come about on its own. It was not an outgrowth of some long, storied tradition of mining as a worker-safety-centered industry. Rather, it evolved—with some speed—as part of a fundamental mindset shift in the industry in the latter half of the twentieth century. In some cases, the shift was particularly rapid. After a dramatic paradigm shift and intensive training program, a certain mining company in South Africa was able to reduce mining fatalities in their operations by 62 percent from 2006 to 2011, and reduced mining fatalities throughout South Africa in general by 25 percent through propagating their techniques.[27] Moreover, the time lost—and therefore the profit lost—by time spent on injuries at every level, from managerial paperwork to worker recuperation, was reduced by more than 50 percent.

A similar mindset shift occurred at Alcoa, the United States' largest aluminum producer, under the leadership of Paul O'Neill. When O'Neill took the helm at Alcoa, its safety record was among the best in the industry—and yet, O'Neill still named safety as his number one goal. Indeed, at his first press conference, O'Neill refused to speak on any issue besides safety. When his investors attempted to steer the conversation toward more conventional topics, such as productivity and stock prices, O'Neill stood firm, reiterating his goal of a company with *zero* workplace accidents.

This zero tolerance for accidents formed the core of O'Neill's commandments and—slowly—he transformed Alcoa's culture from the bottom up. When accidents occurred at individual plants, or when management covered up accidents in order to appear compliant with O'Neill's policy, O'Neill made it a point to personally review the circumstances that led to each safety failure and collaborate with workers, executives, and union reps on policies that would prevent such incidents from occurring again.[28]

O'Neill's efforts paid off. When he first took over as head of Alcoa in 1987, the company was struggling at a valuation of $3 billion. By the time he left in 2000, Alcoa was worth more than $27 billion. This was due in large part to O'Neill's work on safety standards: improving safety improved efficiency, and thus increased revenue.

Moreover, O'Neill's efforts gave the company and its employees a shared purpose. He knew that stock prices, productivity, and the other standard topics of concern among investors would not motivate his employees. And yet he also knew that such a transformative policy could not succeed without his employees' backing. A focus on safety, which connected personally to every employee from the ground level and up, proved a positive, disruptive process. Enthusiasm for such a process would spread organically throughout the entire company.

The implications are obvious: a focus on safety enhances working conditions and directly improves the lives of employees. At the same time, it also increases productivity and, hence, profit. And finally, it helps rally employees around a cause and an identity, which enables all of the above. A true win-win.

Sustainability deserves a similar mindset shift in the mining industry. Just as we perform safety shares as part of our everyday routine—thus inculcating safety as a core value from the top down—so should we incorporate sustainability as a core cultural value. In this culture, producers won't see the environmental and social requirements as a way to maintain their permit-to-operate, or as a trick to convince shareholders, but as part of our diligence to ourselves, our families, and to each other.

This ties into our previous discussion on narrative and awareness. Mining needs to embed a sustainability narrative within its own industry just as much as it needs to project it externally. If it does not practice what it preaches, the mining industry risks presenting an inauthentic face to the outside world. This would be a massive misstep, both in terms of changing mining's public narrative as well as investing in the future of our workforce.

The Sustainability Shift

Sudbury, a relatively small town in northern Ontario, Canada, was once so denuded of vegetation and animal life that NASA used it as a training area for astronauts because its barren nature resembled the airless, lifeless terrain of the moon. An interesting tidbit, albeit bleak. However, I find Sudbury notable for a separate (if related) reason: it is home to some of the largest nickel mines in the world, including one of the few mines that has truly embraced sustainability in recent years.

This was not always so. The copper and nickel smelters in Sudbury were once the single greatest source of sulfur dioxide in the world.[29] Sulfur dioxide is, regrettably, one of the more virulent greenhouse gasses, and thus a significant contributor to human-driven climate change. In large part, this was a relic of what we might call "the bad old days." The nickel and copper mines in Sudbury were constructed in an era before much was known about the environmental impact of mining. Nothing symbolized this prior ignorance more than one of Sudbury's more notable landmarks: the Superstack. A smokestack of cyclopean dimensions, it was constructed in 1972 and measured 1,250 feet at its peak, earning it the dubious honor of being the world's tallest smokestack.

At one point, Sudbury was famous—or rather, infamous—for the environmental devastation it suffered. The landscape surrounding its mines was a testament to human indifference: devoid of vegetation and animal life, it resembled a desert, more a glimpse of some future dystopia than of a northern Ontario town.

But with the help of scientists from Laurentian University such as biologist Graeme Spiers and ecologist Peter Beckett working alongside environmental activists and the local community, Vale (the company that now owns the Sudbury mines, which first operated under the Inco banner) has turned this story of human excess into a parable of modern mining. Scientists at the mine collaborated with specialists from the local university in what they dubbed the "regreening" effort. This involved a specialized lime created to neutralize the pH of the soil surrounding Sudbury, followed by a mass replanting project.

Since the regreening began in 1975, more than 10 million trees and shrubs have been successfully planted in Sudbury. Sudbury, once so barren of life that it was used by NASA as a lunar training area for astronauts, has become one of the great environmental success stories of the last forty years. It possesses some of the cleanest air in Ontario, has successfully regreened more than 100,000 acres of land, and aims to regreen about another 100,000 acres.

Since the regreening began in 1975, more than 10 million trees and shrubs have been successfully planted in Sudbury.

And it did all this while remaining one of the quintessential mining towns in Canada. There are anywhere from seven to eleven active mines at any one time in the city's hinterlands, and the majority of the city's population still finds employment either in mining or mining-support industries. Vale continues the work started by Inco, aiding and abetting the regreening process without reservation.

This story is relevant because it is one of the most immediate and dramatic examples of AIM as a successful strategy for the mining industry. Prior to its purchase by Vale, Inco was not invested in the environmental damage it caused. But after the Canadian government served it an ultimatum—40 percent less greenhouse gas emissions by 1995—Inco took stock.

In a prescient move, Inco came back to the Canadian government and proclaimed it would do the regulations one better: not only were they on board, but they were also going to exceed the mandate. Inco proceeded to collaborate with the local government, activists, and scientists in the restoration and regreening of Sudbury. Inco understood the power of awareness: they did not want a reputation as an environmentally unfriendly company at a time when environmentalism was on the rise, particularly among younger people. By embracing green tech and sustainability and agreeing to a deeper collaboration with local regreening efforts, Inco understood they were investing in the future goodwill of their consumers, their local community, and their future workforce. This mindset shift paid off: Inco made a strong profit off the regreening efforts, sustained Sudbury's long-term viability as a mining town, and earned themselves enough capital to achieve a favorable sale to Vale in the early 2000s.

TYING IT ALL TOGETHER

The AIM framework requires each component to work alongside the others: like gears in a gearset, each contributes its own motivational force to the greater whole. And like that bit of engineering magic, if one piece is removed or malfunctions, the entire machine grinds to a halt. Awareness feeds into investment, and both are enabled only by the mindset shift—yet without either awareness or investment, the mindset shift is meaningless. And arching over them all is sustainability, which provides the organizing principle for the entire framework. It is a methodology that reflects its own values.

> **Incorporating sustainability as a core value of the industry, just as safety once became a core value, is the way forward.**

Incorporating sustainability as a core value of the industry, just as safety once became a core value, is the way forward. The example of Sudbury proves the viability of this methodology. Sustainable mining provides the structure for long-term growth as a business, and the goodwill it garners from local communities is incalculable. When I say sustainable mining, what I mean are those mining practices that minimize the negative impacts—both human and environmental—associated with mining, and which limit extraction rates to a level that does not compromise the potential needs of future generations. Not only will adhering to these principles attract more of the younger generations to our workforce, but it will also attract more investment capital as well.

ESG standards for investment have become *du jour* as of late, driven in large part by the millennial/Zoomer desire for social change. Moreover, data suggests companies with higher ESG scores perform better than—or at least as good as—companies with lower ESG scores, implying that sustainability is a profit-increasing step in the current market, or at least a profit neutral step.

Embracing the AIM framework will provide the mining industry with the tools it needs to survive in this new, sustainability-focused business environment. And due to mining's history, it possesses the tools to accomplish this shift already. This is, in essence, an operational excellence tool. Operational excellence, broadly defined, is a mindset that embraces certain principles or tools to create sustainable improvement within an organization. Operational excellence is achieved when every member of an organization—or in this case, an industry—can see the flow of value to the consumer.

Embracing the AIM framework will provide the mining industry with the tools it needs to survive.

This is a useful mindset for any industry, as the primary objective of operating companies is to maximize production throughput safely while minimizing energy, material consumption, and human costs. In many cases, it is a competitive advantage. In the mining industry, it's a matter of life and death. People, assets, processes, and cash are the foundations of any business. Every other practice and tool supporting operational excellence falls into one of these categories. Sustainability aids us in that it serves all of these categories at once. Our continued operational excellence is therefore inextricable from our link to sustainability. A corporation committed to safe, sustainable, and green business practices will minimize consumption, increase safety (both for workers and outside civilians), and increase cash flow.

The first steps for implementing this AIM framework must be building that awareness and investing in environmentally friendly technologies. These two will be simultaneous and ongoing processes, as they feed directly into one another. Raising awareness by changing the cultural narrative around mining relies on embracing green tech and community relations via collaboration with those local communities, as demonstrated by the Sudbury example. The same processes by which we raise awareness will also constitute our investment in future generations, as embracing green tech and socially conscious values—community collaboration being one—will attract younger generations to mining, and thereby help to build mining's future workforce.

Once awareness and investment have been increased, or at least begun, then comes the mindset shift. Sustainability must be fully developed and internalized as a core value of the mining industry. Just as daily safety shares have become a normalized practice within the industry, so must a daily commitment to greener, more sustainable practices. This goes back to our brief discussion on authenticity: without an authentic, organic belief in sustainability at every level within the operational chain, the mining industry will be unable to sustain any attempts at changing its wider cultural narrative or investing in future generations. This need not be a bleeding-heart commitment to the hugging of trees or a crunchy-granola aesthetic, but rather a recognition that operational excellence is a foundational aspect of continued profitability, and in a modern business, operational excellence is linked at all levels to sustainability.

Sustainability must be fully
developed and internalized as a
core value of the mining industry.

AIM is mining's future. As I said in the introductory chapter, I am proud of my work in the mining industry. What's more, I am proud to continue to work toward a greener, more sustainable industry. I intend to leave future generations with a better world than I inherited; for me, that means focusing on renewable technology. And that renewable technology does not occur without mining, nor can it truly be renewable if mining itself is not green. This is why mining needs AIM.

It is a revolution, but not revolutionary. As evinced by other practices in the industry, as well as success stories like that of the Sudbury experiment, mining has already begun implementing such processes. More specifically, it has implemented aspects of this framework throughout its history, although not the framework as a whole. AIM pulls those disparate elements together, illuminating the trails blazed by other pioneers in the industry and giving mining the tools to future-proof its own business.

CONCLUSION

U ltimately, no matter our age, gender, or nationality, we all have our parts to play in tomorrow's energy needs. Achieving net-zero carbon emissions may seem a difficult goal. Mining's stake in this goal is prodigious, and while not all of us may work in the mining industry—and even those of us who do may, at times, feel unable to influence the industry—all of us benefit from its fruits just as much as we suffer its faults. Therefore, we must all take an active part in the future of mining.

For those in a position to directly influence the mining industry, the path is clear: embrace the sustainable practices outlined above and prepare their share of the industry for a greener future.

For the rest of us, a different path presents itself: pressure. Exerting pressure on the industry as its consumers—and we all are its consumers, of that there can be no doubt—is the strongest lever by which we might shift the industry's weight and start it rolling down the road to a greener future.

After all, the most viable pressure to place upon an industry involves its bottom line: affecting its profits, in other words. For mining, this seems like it might present problems. The vast majority of us do not—indeed, cannot—purchase minerals directly from mining companies. That's just not how the industry functions. And boycotting is, perhaps, not viable: asking people to reject purchasing all those amenities they enjoy that incorporate mined materials is simply an untenable proposition in this digital age.

But there is another way. Just as the mining industry must cultivate awareness of its better aspects among the general population, so must the general population make its own interests visible. For example, take the nutritional information on the side of a soft drink can: it conveys the caloric content, the amount of sugar added, and the ingredients used. As consumers, we can use this information to make informed decisions about which products we consume, and which products we do not.

Imagine then if the same information were available for our everyday products made with extracted minerals: our cellphones, laptops, perhaps even our cars. Imagine if the next time we purchased a television, we could scan a QR code on its side and learn what metals were used in its manufacture, where they were mined, how they were mined, the resultant carbon footprint, and all the other aspects of its journey from the mine to us.

I cannot promise this information will be easy to find, at least not at first. But it was not easy to find this information on the food and beverage products one consumed either, at least until consumers started asking for it and favoring those companies who did provide it.

We must do the same for mining. The more we ask questions, the more the industry must respond. If we can secure this chain of information, we can make informed decisions about which companies to support, and we can choose those with lower emissions and more ethical mining practices. And the more we purchase goods and services from those brands that align with our values, the more the industry as a whole will respond. By requiring certain information be made visible and adjusting our purchasing habits accordingly, we can make our own desires for the mining industry heard. Once we have this visibility, I promise you it will change the way the entire mining industry operates.

Mining is an integral part of our lives and it's not going away. It's also a part of our transition to a clean future. The industry needs to embrace sustainability as a mindset, and we—the consumers—are the catalyst that will speed up that process.

In mining, we do not just dig a hole in the ground. We help build the future of the world. Let's do it sustainably together.

Alp Bora

ENDNOTES

1 Delivinge, Lindsay, Will Glazener, Liesbet Gregoir, and Kimberly Henderson, "Climate Risk and Decarbonization: what every mining CEO needs to know," McKinsey, January 28, 2020. https://www.mckinsey.com/business-functions/sustainability/our-insights/climate-risk-and-decarbonization-what-every-mining-ceo-needs-to-know

2 "Minerals used in electric cars compared to conventional cars," International Energy Agency, May 4, 2021. https://www.iea.org/data-and-statistics/charts/minerals-used-in-electric-cars-compared-to-conventional-cars

3 Rodriguez, Laura, "Rare metals in the photovoltaic industry," RatedPower.com, September 24, 2021. https://ratedpower.com/blog/rare-metals-photovoltaic/

4 E. M. Harper, Zhouwei Diao, Stefania Panousi, Philip Nuss, Matthew J. Eckelman, T. E. Graedel, "The Criticality of Four Nuclear Energy Metals," Resources, Conservation and Recycling, Volume 95, 2015.

5 "Miner Demographics and Statistics in the USA," accessed July 1, 2022, https://www.zippia.com/miner-jobs/demographics/

6 "Mine 2022: A Critical Transition," PWC.com, accessed June 6, 2022, https://www.pwc.com/gx/en/industries/energy-utilities-resources/publications/mine.html

7 "Sustainability," The United Nations, accessed July 1, 2022, https://www.un.org/en/academic-impact/sustainability

8 "Why Mining Is Important and What Is the Impact?" accessed July 1, 2022, https://www.scotforge.com/Blog/why-is-mining-important-and-what-is-the-impact

9 Zwicky, Arnold, "Why Are We So Illuded?" Stanford University, September 2006.

10 West, Larry, "The Benefits of Aluminum Recycling," Treehugger, July 24, 2019, https://www.treehugger.com/the-benefits-of-aluminum-recycling-1204138

11 "The New York State Returnable Container Act," New York State Department of Environmental Conservation, retrieved August 22, 2022. https://www.dec.ny.gov/chemical/8833.html

12 "Our Commitment to a Goal of Net Zero by 2050 or Sooner," International Council of Minerals and Metals, October 5, 2021, https://www.icmm.com/en-gb/

our-work/environmental-resilience/climate-change/
net-zero-commitment

13 "Scope 1 and Scope 2 Inventory Guidance," Environmental Protection Agency, September 29, 2021, https://www.epa.gov/climateleadership/scope-1-and-scope-2-inventory-guidance

14 "Scope 3 Inventory Guidance," Environmental Protection Agency, September 29, 2021, https://www.epa.gov/climateleadership/scope-3-inventory-guidance

15 "Miner Demographics and Statistics in the USA," accessed July 1, 2022, https://www.zippia.com/miner-jobs/demographics/

16 E. Napoletano, "Environmental, Social and Governance: What Is ESG Investing?" Forbes Advisor, Forbes.com, February 24, 2022, https://www.forbes.com/advisor/investing/esg-investing/

17 Alana Benson, "Environmental, Social and Governance (ESG) Investing and How to Get Started," nerdwallet.com, June 23, 2022, https://www.nerdwallet.com/article/investing/esg-investing

18 Nina McQueen, "Workplace Culture Trends: The Key to Hiring (and Keeping) Top Talent in 2018," LinkedIn.com, June 26, 2018, https://blog.linkedin.com/2018/june/26/workplace-culture-trends-the-key-to-hiring-and-keeping-top-talent

19 "A kilo more of copper," Renewable Energy Magazine, April 14, 2021. https://www.renewableenergymagazine. com/energy_saving/a-kilo-more-of-copper-increases-environmental

20 "20 Staggering E-Waste Facts in 2021," Earth911, October 11, 2021, https://earth911.com/ eco-tech/20-e-waste-facts/

21 Ugo Bardi, Extracted: How the Quest for Mineral Wealth is Plundering the Planet, Chelsea Green Publishing, 2014.

22 "Electric Donation and Recycling," Environmental Protection Agency, accessed August 22, 2022, https:// www.epa.gov/recycle/electronics-donation-and-recycling

23 Charmaine Crutchfield, "Smartphone Disposal Poses Security Risks, Experts Warn," USA Today, November 10, 2014, https://www. usatoday.com/story/news/nation/2014/11/10/ smart-phone-security-risks/18798709

24 Lindsey Gilpin, "The Depressing Truth About E-waste." TechRepublic, June 11, 2014, https://www.techrepublic.com/article/ the-depressing-truth-about-e-waste-10-things-to-know

25 C. P. Balde, V. Forti, V. Gray, R. Kuehr, P. Stegmann, The Global E-Waste Monitor 2017: Quantities, Flows, and Resources," United Nations University, 2017.

26   "The Deloitte Global 2022 Gen Z and Millennial Survey," Deloitte, accessed August 22, 2022, https://www2.deloitte.com/content/dam/Deloitte/global/Documents/deloitte-2022-genz-millennial-survey.pdf

27   Cynthia Carroll, "Moving to a Safety Culture in Mining," Harvard Business Review, January 15, 2014, https://hbr.org/2014/01/moving-to-a-safety-culture-in-mining

28   David Burkus, "How Paul O'Neill Fought for Safety at Alcoa," April 20, 2022, https://davidburkus.com/2020/04/how-paul-oneill-fought-for-safety-at-alcoa

29   Sara Miller Llana, "The Sudbury Model: How One of the World's Major Polluters Went Green," The Christian Science Monitor, September 24, 2020, https://www.csmonitor.com/Environment/2020/0924/The-Sudbury-model-How-one-of-the-world-s-major-polluters-went-green

www.ingramcontent.com/pod-product-compliance
Lightning Source LLC
Chambersburg PA
CBHW030904180526
45163CB00004B/1703